nF418640

Introductions

First of all, I would like to congratulate you for purchasing what might be the next technological breakthrough of the 21ˢᵗ century: the Apple watch. What could be known as the next revolutionary marvel in terms of technology, the Apple Watch is not a simple time-piece. In fact, you'd be in for a treat for the ways it can help you in your everyday life. Consider, for example, that there is an accessory so small, simple, yet bewilderingly innovative that it can help you with almost every day activity. Seems far-fetched? Not so much as you think, since it's already here now.

In fact, you'll be amazed that it can do more than what you've expected. Stay in touch, in shape, and in sync with the all new Apple Watch. With this revolutionary technology that Apple brings, this accessory fitted for the wrist is also bound to suit your personal style with its array of designs. Truly, this piece is an impressive addition to any tech household.

Apple Watch, to date, is a truly unmatched piece of convenient accessory that caters to any tech-loving person. Thanks to its number of apps that monitor your body, namely, built-inhear rate sensors, regular pulse monitors, accelerometers, etc., they can keep track of the steps you take and speed, together with the advantage of allowing your iPhone's GPS to measure distance. Aside from how you can keep track of your health, the Apple Watch also serves as an excellent communication tool. With it, you can now immediately respond to messages and phone calls, as well as share heartbeat and sketches with significant others using their Apple Watch. Truly innovative, they have now surpassed the point where it can replace your wallet when making purchases. Thanks to Apple's Apple Pay, you can now present it when travelling without the need of digging around for your boarding passes or tickets. Lastly, the Apple Watch can serve as your ultimate guide by keeping you informed of your location, the weather, stock's performances, and many more! And, similar to most all of Apple products, built-in apps is just the start. Third party developers are widely extending the functionality of the Apple Watch by constantly creating a stream of widely-developed apps—without a sight nearing the end of it.

Apart from their seemingly practical applications, the Apple Watch also has the significant power to alter the course of our relationship to technology. By bringing our digital life closer than ever before, they've managed place us with technology by constantly putting us in constant physical contact with it. Surely, the Apple Watch cannot replace your iPhone but rather aids you in managing it well and significantly reducing the time you spend staring at it.

In this guide, we aim to provide with everything you need to know about the Apple Watch. Aside from providing you with content readily found in the user guide of the watch, we'll delve deeper into more advanced information not readily found in the web. Since obviously you can easily find most features from the Apple website and official manuals, we'll extend it further by exploring every nook and cranny of the Apple Watch.

Despite its small size though, there is a surprising number of things hidden away in your little gem of an Apple Watch. Towards the first chapter, **Part I: An Intro For The Apple Watch,** we'll show you exactly what the Apple Watch is under the hood. In **Part II: Why do we even need it?** We'll highlight how the Apple Watch can serve as an everyday accessory and not as a luxury item. In **Part III: Apple Watch Basics,** we'll guide you on how to set it up and navigate it. Towards **Part IV: Personalizing your Apple**

Watch, we'll guide you on how you can customize the way you use your Apple Watch and how it responds back. In **Part V: Apple Watch's Basic Apps,** we'll explore all the built-in apps that comes with your watch and how you tweak it. Then, in **Part VI: Managing Third Party Apps,** we'll share with you some of the most loved third party apps that can be useful for everyday use. And lastly, in **Part VI: Apple Watch Alternatives,** we'll suggest other similar accessories close to what the Apple Watch is capable of. Though there may be more to the Apple Watch that is still yet unknown to most, we hope to find that you'll enjoy this beautiful piece of gem and will likely find more discoveries on how it can make your life easier.

Chapter 1: Introduction of the Apple Watch

The Apple Watch, is one of the most awaited product from the impressive lines of Apple- in fact, it is the first real new product released since Apple introduced the iPad back in 2010. So, some introductions must be in order before you get started with this entirely exciting, exciting new product from Apple.

First things first, the Apple Watch is Apple's first "wearable" device. Wearables, though fairly young in the tech family, are will soon be a commercially viable branch thanks to smart watches. Of course, it shouldn't be forgotten that the ancestors of the modern wearables include plain old pedometers, which could count steps, distance travelled, and calories, and times can even detect elevation changes. In recent years, dedicated fitness trackers from FitBit, Jawbone, Garmin and Pebble have moved way beyond step counts by sending their data to smartphone apps and online platforms, giving users ways to visualize and use that data to set goals and improve performance. These companies also pioneered the first steps toward a full-featured "smart watch" by adding things like call and text notification.

Apple Watch is not the first smart watch out there. The Android Wear platform has already released seven Android-based smart watches, with more in the works. Microsoft is looking to get into the wrist game as well, with the Microsoft Band being released in 2015. So why choose Apple Watch? It's not the cheapest option, nor is it the only option. Still, if you are an iPhone user, Apple's hardware is gorgeous, its software is powerful, and it builds on the interface that you already know and use.

If you're interested in fitness tracking, Apple Watch can compete with the big names, and thanks to third party apps, users can greatly extend its utility past the out-of-the-box configuration. If you want to wear something you'll like looking at, the Apple Watch comes in several different models, including the rough-and-tumble Watch Sport (with an aluminum case and a sweat-proof plastic band), Watch (with a stainless steel case and your choice of a variety of band styles) and, for high rollers, Watch Edition (with a gold case), which can run you up to $15,000. You can also purchase bands separately, meaning you can take advantage of the comfortable and water-resistant Sport band when you're working out and still wear something a little more elegant when out on the town. In short, it's a highly personalized and highly sophisticated device with excellent features and revolutionary technology. What's not to love?

Apple Watch: What's it made of?

The technology inside the Apple Watch is an impressive achievement in and of itself. If you were to open up the Watch's case (note: don't actually do this!), you'd find a single chip inside. That chip – named the S1 Processor - contains an entire computer system, and it's the first of its kind. It's particularly awe-inspiring when you remember that the earliest computers took up entire rooms!

The Watch case also includes the all-new Taptic Engine. This piece of hardware provides haptic feedback, meaning it taps and buzzes on your wrist to let you know that something's up. On the back of the device, you'll find its heart rate sensor, which uses infrared and LED lights to monitor and record your pulse.

Finally, your Apple Watch includes an incredibly strong display. Apple Watch and Apple Watch Edition models feature a sapphire glass display (which, incidentally, is second only to diamond in durability). The less expensive Sport model's display is made of strengthened Ion-X glass, which is also tough enough to withstand the demands of your daily life.

Apple Watch and iPhone: The Dynamic Duo

Before we get too much further, we should be clear about something. You *must* have an iPhone, or Apple Watch will be pretty useless. The Apple Watch is designed to work with an Apple iPhone 5 or later. Unfortunately, anything older than iPhone 5 can't keep up with the Watch in terms of technology. Furthermore, your iPhone needs to be running iOS 8.2 or later. The iOS 8.2 update adds a dedicated Apple Watch app to the iOS interface, and as an Apple Watch owner, you'll be using that app to control and customize your Watch. If you don't have it yet, open the Settings app on your phone and then tap **General**, then **Software Update** to update to the latest version of iOS.

Apple Watch talks to your iPhone through Bluetooth – a wireless technology found in cars, headphones, wireless keyboards and more. That connection makes it possible for Apple Watch to send messages, stream media and give directions. Your iPhone shares its wireless or data connection with the Watch, which can't connect to either on its own. However, Apple Watch *can* send messages and receive and make phone calls if it is connected to the same wireless network as its paired iPhone, but many other features won't be available. Long story short, you'll want your iPhone within 30 feet or so (the standard Bluetooth range) of your Watch to get the most out of it.

Getting Started!

Now that you know what you're working with, read on to get familiar with Apple Watch basics by setting up your Watch and learning to navigate it.

Chapter 2: Why Do We Even Need it?

In this chapter, we will discuss how important the Apple Watch really is; how it's an everyday accessory that can be worn by most rather as a luxury that isn't really necessary for everyday use.

As you may notice, being in the 21st century, we can consider watches have been out for decades. They just aren't the hype any more since the dawn of the cell phone. Ever since cell phones was introduced to the mass populace, watches faced a steady decline as the everyday piece of accessory everyone needed. In fact, almost no one in grade school has a watch any more. Instead, all the kids are given cell phones—smart phones, to be exact. Aside from that, they also have iPads and iPod touch (ridiculous, we know.) But really, who would still ask for a watch for Christmas?

The million dollar question comes now: Do we even need the Apple Watch?

To be blunt, we don't necessarily NEED the Apple Watch. But, come to think of it, there are other things that we do not NEED but feel like such a necessity is required for everyday living. If, then, come to think of it, when you already have a laptop, what purpose that does tablet serve? If it's really the touch display, shouldn't your phone be enough? Or maybe it's rather a more convenient way of accessing the two?

That being said, one way of thinking what the Apple Watch's purpose really is, is how it can be the next step towards the connection of people and technology. It's widely evident that people and technology had intertwined and found comfort in each other ever since the age of smartphones. The way they interact with their tablets, read books on their Kindles, etc., is no way near how people used to interact with their computers in, for example, the 1970s.

Connections between people and devices were apparent back then. However, decades ago, they weren't that much connection between the two. Back then, computers were bulky, slow, and would require much effort to be used. Up until the introduction of laptops were we only able to be brought closer to technological experience. As laptops grew smaller, our connection with them significantly increased and their convenience had made us bring them to coffee shops and airports.

After which, the tablet was introduced. It gave us the similar power of the laptop only in a smaller form. As they've developed it, they managed to fit the cell phone and the likes of a laptop unto an even smaller scale: the smartphone. Since then, our phones had always been inside our pockets together with knowledge and power they possess.

If all of this had been apparently done in order for us to be more technologically advanced on a smaller and smaller scale, then shouldn't our wrist be the next place for advancement?

Despite all this, our interactions with the computer haven't changed that much. We sit in front of the computer, stare at the screen, and then make a conscious effort to break from reality and unto the digital world through our computer.

As we are drawn closer technology, our attention is immediately lost on our device. Take this for example. When you pull out your phone from your pocket or purse to check a message, swipe it or unlock it, then suddenly, BAM! Your full attention had been fully diverted unto your phone—even after your only intent of checking the message. You are suddenly tempted to use other apps while you wait for a reply or to simply kill time. Before you know it, you've already spent hours and hours using it unintentionally.

Nowadays, you can easily look around and spot people using their phones. Sure, they're physically present—walking, talking to each other, eating, etc. —however, they're not all mentally focused. Their attention is trapped in their world of phone. Another thing with it, is that it can cause a myriad of distractions that can ultimately result to an accident. Say you glance at your phone while driving, suddenly you are caught up with a notification about a friend, you check your Facebook page and unknowingly steer into who-knows-what. Even checking your phone for a minute may result to the temptation of using it longer, which can pose as an immediate threat.

With this in mind, the Apple Watch can come into play. So, how can we now justify that we need it?

Information can now be presented to us with a glance at the wrist, which doesn't really take up much of our attention since that is the extent of it: a glance. Using pre-selected options, you can easily reply to texts without diverting your attention from the road. What's more, it can be done with a simple tap. An ease and speed of interaction is present with the Apple Watch that is not available with our phones.

This convenience can help us be back in the physical world, which is more valuable than what you might think. The Apple Watch is one way to increase your engagement with your surroundings whilst engaging and keeping up with current events through technology.

From observation, people who are currently enjoying their Apple Watch or other smart phones tend to:

- Take their phones out less frequently.
- Looks at their wrists more frequently.
- Enjoy that instant access to information
- Enjoy the apps on the watch
- Enjoys being on-the-go while being updated.

Come to think of it, the Apple Watch is a necessary body piece that can improve one's life. As you've probably experienced during the past decade of your life, the most popular items that you've bought are items that helped you saved time. This, then, allows you to spend more time on things that really matter.

It's not a luxury watch—it's an everyday item.

Many people are trying to analyse the Apple Watch with the assumption that the device is just another luxury watch only with additional customization and features. I think this type of thinking misses the big picture. An Apple Watch will be just as much a watch as an iPhone is a phone.

By keeping the message simple, anxiety and uneasiness will be removed from the buying decision. Most consumers, even if they don't wear a watch, understand what a watch is and what it does. However, the comparison of Apple Watch to a luxury watch needs to stop there as once a user begins to rely on Apple Watch for communication, health and fitness tracking, and mobile payments, the idea that it is just another luxury watch will no longer apply. Some people think it also will bring back wearing wristwatches into common practice. Watch-wearing fell off sharply after cell phones started doing double duty as timepieces. As most millennial quip, "Watches are so 20th century."

The Apple Watch and its strengths shouldn't be compared to luxury watches, and more importantly, luxury watch strengths. Timelessness or lack thereof, seems to be at the top of the list of lingering questions about Apple Watch. If a luxury watch can last the test of time and be passed down from generation to generation, how would Apple Watch compete? Who would pay thousands of dollars for a device that won't stand the test of time? Timelessness won't matter for Apple Watch since the Apple Watch isn't a luxury watch. Instead, Apple Watch is a mobile computing facilitator worn on the wrist. The users will have just as much motive and desire to pass the device down to children or family as they would with an iPhone or iPad. By discussing price in context of luxury watches, I suspect many are jumping to the conclusion that the only reason someone will pay thousands of dollars for an Apple Watch is to wear it forever as a status symbol. Instead, people will pay thousands of dollars in order to have the opportunity to buy an Apple product that can be worn. The desire to upgrade to a newer, more advanced version in the future will likely be just as strong as it is with iPhone.

Apple understands its user base very well and correctly sees that there is a market for very high-end tech gadgets. This buyer takes an iPhone and its lack of personalization, puts thousands of dollars into the device to truly make it his or her own, and then will eventually upgrade to a newer iPhone just like everyone else. The Apple Watch Edition collection is Apple's first attempt at addressing this segment of the market.

Just as with fashion, technology evolves. The Apple Watch isn't a luxury watch, but rather a fashionable communication facilitator worn on the wrist.

Chapter 3: Apple Watch: Basics

In this chapter, we'll take you from zero to sixty with your Apple Watch by showing you everything you need to know to work it. This section includes all of the basics – unboxing, charging, pairing and navigating, as well as an introduction to the major elements of the interface and of the iPhone companion app. Take your time with this section. Apple Watch will require you to learn some new behaviors, so be patient with yourself as you acclimatize to this new style of mobile computing.

3.1 A First Look of the Apple Watch

When you first open your Apple Watch box, you'll find the Watch itself inside a plastic case. Underneath that case, you'll find a long envelope with some minimal use instructions and safety information, a USB charger cable with the Watch's new magnetic charger on one end, and an AC power adaptor. Depending on your band purchase, you may also find a band adaptor for larger wrist sizes.

Gently remove the protective plastic from your Watch before putting it on. Many Apple Watch bands, including the Sport models, have a slightly different clasping mechanism than most wristwatches. You'll feed the perforated end down into the slot in the end with the snap, which will fit snugly inside its hole. However the Watch buckles or clasps, it should feel secure and comfortable on your wrist.

3.2 The Charging Process

The Apple Watch comes with a magnetic charger, USB cable and power adaptor. Plug the adaptor into a power source and connect the USB cable. Your Watch will sit on top of the circular magnet as it charges. Your Apple Watch should ship with enough of a charge for you to get started, but you'll want to figure out a workable charging routine pretty quickly, since charging will generally be a nightly task.

Your Watch will hold a charge for about a day's worth of normal use. You may see longer or shorter battery life, depending on how you use your Watch. Apple claims that the battery will last for three hours of talking, six and a half hours of streaming music, and six and a half hours of fitness tracking. By putting your Watch in Power Reserve mode, you can expect to be able to check the time for up to 48 hours. Expect a full charging cycle to take about two and a half hours.

When your Apple Watch is charging, you'll see a small green lightning bolt on your Watch Face, indicating that it's charging, as seen below. Then, turn on your Apple Watch by pressing the slender silver button next to the digital crown along the right edge of the Watch. The Apple logo will appear. It may take several minutes for the initial setup options to appear.

When the display shows your language options, choose your language by tapping it. English is at the top of the list, but if you want to see how the knob on the side of the Watch called the Digital Crown works, twist it to scroll through the rest of the options. Apple Watch will then ask you to pair it with your iPhone. Tap Start Pairing to reveal the graphic containing everything your iPhone needs to know.

3.3 Paired up: Apple Watch & iPhone

Now that you're familiar with the hardware of your Apple Watch, it's time to start using the software it supports! We'll walk you through the process of pairing your Apple Watch and your iPhone so that you can get started. To complete this setup process, you'll need your Watch, an iPhone 5 or later, a reliable wireless or data connection, your Apple ID and password, and five to ten minutes of your time.

First, open the Apple Watch app on your iPhone. You'll need to be updated to iOS 8.2 or later to have this app, and you must have it installed in order to use an Apple Watch.

Next, on your iPhone, tap Start Pairing in the Apple Watch app. You can pair your Apple Watch automatically or manually. The more fun method is the first one presented to you – simply line up your Watch with the square on the iPhone, and the two devices will pair using Bluetooth, courtesy of the animated identifying graphic on the Watch.

If you're having trouble with this method, you can tap Pair Apple Watch Manually at the bottom of the screen. You'll then choose your Apple Watch from the list of available devices. In most circumstances, we imagine this will be a fairly short list. If you're unsure though, tap the little circled "I" in the upper right corner of your Watch to find your device's name. Once you've selected it on your iPhone, a six-digit code will appear on your Watch. Type that code into your iPhone to complete the pairing process.

 You'll then see an iPhone confirmation screen letting you know that the two devices are happily talking to each other.

Now, you'll use your iPhone to finish setting up your Watch. Tap Set Up Apple Watch to continue.

First, choose your wrist preference by tapping Left or Right (you can change this later if you need to). Then, tap Agree, indicating you've read and agreed to the terms and conditions. Tap Agree again in the popup confirmation box. Then, sign in with the Apple ID associated with your iPhone. You can skip this step if you want, but we strongly recommend signing in. An Apple ID is required for some of Apple Watch's more exciting features, including Apple Pay. Enter your password and tap Next in the top right corner.

Next, you'll see information about Location Services and Siri. These features are enabled or disabled based on your current iPhone settings. If Location Services are off on your iPhone, they'll be off on your Apple Watch as well. These particular features really enhance the Apple Watch experience, and we do recommend enabling them on your iPhone if necessary. Tap OK to confirm you understand each setting.

Next, decide whether or not to send diagnostic and usage data to Apple by tapping either Automatically Send or Don't Send. This one, unlike the Terms and Conditions agreement, is entirely up to you!

After that, you'll be given the option of entering a passcode. A passcode is a four-digit code that you'll need to enter if you take off your Watch, which will lock whenever it's removed. If you're going to use Apple Pay, you'll be required to set up a passcode in order to access the feature. However, you can tap Don't Add Passcode if you don't plan to use Apple Pay and don't think you need that level of protection. If you're on the other end of the security anxiety spectrum, though, you can tap Add a Long Passcode to use a longer string of numbers.

The next screen asks you if you'd like to install apps on your iPhone that are compatible with Apple Watch. We do recommend selecting Choose Later at this point. Apple Watch does have a little bit of a learning curve, and we suggest getting used to the interface and navigation before adding more elements to it.

The final step of setup is a syncing process that may take a few minutes. You can pass the time by tapping See How Apple Watch Works to view a short video explaining the Watch if you like. Your Apple Watch will tap you on the wrist as soon as it's finished. You'll receive a message on your iPhone (and all of your other iDevices) that a new device is using your Apple ID.

When your Watch is ready, the iPhone app will display a confirmation screen, and your Apple Watch will display its default Watch Face (Modular View), which will look something like the screenshot below.

At this point, tap OK on your iPhone to reveal the Apple Watch management features of the iPhone app. We'll be coming back to this screen later on, but for now, let's start learning our way around your Watch!

3.4 Apple Watch Navigation

In this section, we'll show you how to interact with your Watch and clue you in as to how your Watch will interact with you. The Apple Watch interface is a little less intuitive than that of iOS. We found it took us a couple of days to really get the hang of it, so don't worry if you're not immediately up on your feet.

Longtime iOS users especially may need to spend some time getting used to Apple Watch's controls. An iPhone is almost entirely controllable through the touch screen, with the Home button being about the only exception. The Apple Watch is a touch screen device, but the smaller size of the screen effectively limits your precision when tapping and swiping. Here's how the Apple Watch compensates.

Digital Crown

The Apple Watch's Digital Crown feels like a throwback to the scroll wheels of early iPods. However, it makes quite a bit of sense as an alternative to pinching to zoom or using a finger to scroll. The Digital Crown is the knob located on the right edge of the Watch, where you'd expect to find a setting mechanism on a traditional wristwatch. Twist it to zoom in on a screen, scroll through a list, or select an option.

Side Button

The Side Button (located next to the Digital Crown on the right edge of the device) picks up a few loose end features. Perhaps most obviously, it serves as a power button. Press and hold it to power Apple Watch on or off. Pressing it briefly will show or hide your Friends screen. It's also important for Apple Pay – tap it twice to activate that feature.

Home Button

The Digital Crown also functions a button – you can press the whole thing down. This serves as your Home button. You'll use it to return to your Home screen or to your Watch Face. To move quickly between recently opened apps, simply double tap the Home button.

Heart Rate Sensor

Apple Watch's built-in heart rate monitor is located on the back of the device and can quickly take your pulse for you and store the information in Apple Health

Microphone and Speaker

Your Watch can hear you and talk to you, courtesy of the tiny microphone and speaker located on the back of the device.

Touch and Force Touch

Apple Watch adds another dimension to the now-familiar touch screen repertoire. Apple Watch actually registers force in addition to position and duration of your taps through something Apple calls Force Touch. Forcefully pressing your Watch will allow you to access options and features for many apps – a little bit like right clicking on a computer. We found that Force

Touch required really forceful tapping, so don't be afraid to poke your device like you mean it – we promise you won't break it!

.

Haptic Feedback

Haptic means related to touch, and Apple Watch's Taptic Engine allows your Watch to tap back. The Taptic Engine is located on the back of the Watch and sits flush against your wrist. An incoming notification will cause your Watch to give you a nudge to get your attention.

Siri

Siri, Apple's famous digital assistant, is built into the Apple Watch. Even if you've never been a huge Siri user, we think you'll be converted, at least where your Watch is concerned. Since an onscreen keyboard would be highly impractical on such a small device, Siri takes over as your primary means of text entry. She can help you send messages, search for locations, set up calendar events and more. On the Apple Watch, she's especially helpful when it comes to opening apps, which are represented by tiny circles. Once they start accumulating, it may be easier to ask for them than to poke around your Watch.

To access Siri on Apple Watch, simply raise your Watch-bearing wrist toward your face and say, "Hey Siri!" This will activate Siri. Speak your command, and Siri will work her magic. Of course, if you're a bit embarrassed by shouting at your Watch in public, you can also press and hold the Digital Crown to activate Siri with a little more dignity.

By now, Siri is a pretty smart lady. She can understand a fairly large range of inquiries. Here are some common ways to ask Siri to help out on your Watch:

- "Open Maps." (launches the Maps app)
- "Create a calendar event for Team Meeting on June 2 at 3:00." (adds an event to your calendar)
- "Remind me to pick up milk." (creates a reminder on your iPhone)
- "Play Tom Waits." (plays music from your iPhone)
- "Tell Chris I'm going to be home late." (sends a text message)

For more suggestions, activate Siri on your iPhone by pressing and holding the Home button. Then, tap the question mark in the lower left corner.

3.5 Apple Watch: Interface

Now that you know how to get around, let's look at all the places you can go. On your Apple Watch, you'll need to be familiar with your Watch Face, Home screen, Glances, Friends and Notifications in order to feel comfortable getting around. Let's take a look at each of these elements one at a time.

Home Screen

The Home screen is quite similar to that of an iPhone, though instead of a neat grid, your apps are arranged in a circular cloud. To open an app, tap it, or use your finger to move around the app cloud and then use the Digital Crown to zoom in. If that proves a little too difficult, ask Siri to help ("Hey Siri, open Maps"). You can rearrange this screen just like you can on your iPhone. Tap and hold an app until they all start shaking. Then you can drag them around into your preferred configuration. Of course, we think it's much easier to do on your iPhone's bigger screen. To edit your layout on your iPhone, open the Apple Watch iPhone app and tap App Layout.

Watch Face

When you look at your Apple Watch, the first thing you'll see is your Watch Face. This is customizable, but straight out of the box, you'll see the "Modular View" of your Watch Face. This includes the time, date, upcoming calendar events, weather, Activity information and your world clock. You can tap on each piece of information to open it in its respective Watch app (for example, tap the three concentric circles to launch Activity).

From here, you can do a few other things. Press the Digital Crown to view your Home screen, which we'll talk about next. Swipe down to view Notifications. Swipe up with your finger to open your Glances.

Notifications

Your Apple Watch will notify you as new stuff comes in, whether it's a new text message, a Facebook Messenger message, a reminder to stand up and move around, or many other possibilities. Your Watch will sound a tone and tap you on the wrist to let you know about an incoming notification. You'll also see a red dot appear at the top of your Watch Face, as shown below.

To view your notifications, swipe down from the top of your Watch Face. You can dismiss notifications on your Watch or on your iPhone. The screenshot below is an example of a notification from an app on our iPhone. Breeze, at the time of writing, is not a Watch app, but any app that displays notifications on your iPhone can notify you on your Watch as well. This can get overwhelming quickly.

Many Apple Watch app notifications can be handled as they come in. For example, you can respond to text messages and calendar invites from within the notification screen on your Apple Watch. To dismiss a notification, tap Dismiss at the bottom of the notification.

Friends

Pressing the Side Button reveals a list of up to twelve of your

friends. You can make calls or send text messages from your Watch, and if your friends are lucky enough to have Apple Watches, there are some other fun ways to stay (nearly literally) in touch with each other.

3.6 iPhone's Apple Watch App

The Apple Watch app on your iPhone is a key player in your Apple Watch ownership experience, and you'll want to familiarize yourself with it sooner rather than later. You'll use this app to customize your Watch's layout and many of its settings, as well as discover and install Watch-friendly apps.

The Watch app includes four major sections, which you'll find at the bottom of the screen – My Watch, Explore, Featured and Search. The first section – My Watch – is where you can manage all of your Apple Watch settings. We'll be visiting this screen throughout this guide.

The Explore section includes several Apple Watch promotional and tutorial videos. We found these videos to be a bit heavy on marketing and bit light on instruction, but they're well produced and will make you feel quite good about your purchase!

The Featured section is a bit more practical. This screen is actually a Watch-friendly subsection of the Apple App Store, where you'll find a curate lists of the best third party Apple Watch apps.

Finally, the Search screen will allow you to search for a specific Watch-compatible app.

3.7 Working with Two Devices

Apple Watch can send things like Mail messages over to your iPhone. This is a great way to work between two devices. When you open a calendar event, text message, email or any number of pieces of content on your Apple Watch, a small icon will appear in the lower left corner of your iPhone's lock screen. Tap that icon to go directly to that item on your iPhone for editing or closer reading.

3.8 Getting the hang of it.

Apple Watch may feel a little strange at first. We promise though that you'll get the hang of it after a day or two of use. The best way to get used to the different screens and buttons is to play with them. Eventually, you'll get into the rhythm. Take your time and have fun learning your device, and it will be second nature before you know it! The next order of business is truly personalizing your Watch. Read on to learn how.

Chapter 4: Customizing Your Apple Watch

Apple Watch is designed to be an incredibly personal device. Consequently, it works best when it's customized to your own needs. In this chapter, we'll show you how to truly make your Apple Watch your own, through customizing your Watch Face, managing your notifications, adding your friends to the Friends screen and communicating using Digital Touch, and changing out your Watch's bands.

4.1 Different Watch Faces of Apple Watch

To start, let's talk about the most watch-like aspect of the Apple Watch – the Watch Face. Your Watch Face displays every time you flick your wrist toward your face, using the natural and familiar gesture of checking a normal wristwatch. It's a brilliant design that allows this expensive and sophisticated piece of tech to function just like a standard wristwatch, in addition to everything else it does!

Of course, there's more to the Watch Face than just telling time. Your Watch Face is the most customizable feature of your Watch, and it's probably the one you'll look at the most often.

At the time of writing, the Apple Watch includes ten different styles of Watch

Face – Modular (the default), Simple, Motion, Astronomy, Color, Solar, Chronograph, Mickey (sure to generate a fair bit of nostalgia!), X-Large and Utility. Watch software updates will likely include additions to this list, and while Apple currently does not allow third party Watch Face development, the company has hinted that it might be possible in the future.

Most of Apple's styles are customizable in terms of their look and feel and in terms of the information displayed. We highly recommend considering which Watch Face style will work the best with your own sense of personal style and with your own unique utilization of the Watch itself. If you can't decide on just one, don't worry – it's easy to switch back and forth!

To change your Watch Face style, Force Touch the display while the Watch Face is showing. Swipe left and right to see all of the available styles.

Modular

The Modular Watch Face is fairly straightforward. It includes the time (of course) and has space for five widget-like app connections. This means that the Modular Watch Face can give you a pretty large amount of information in a single glance, without having to touch or adjust your Watch at all. The four smaller squares can be customized to display the date and day of the week, the current moon phase, the time of today's sunrise/sunset, the current temperature, stocks information, your Activity stats, the Apple Watch alarm, the Apple Watch timer, the stopwatch, your remaining battery, or each World Clock location you've entered into the World Clock app on your iPhone.

To customize Modular, Force Touch the display while the Watch Face is showing. This will bring up the customization button. There are two customization screens available – color and widget selection. Swipe left and right to move between these two screens. To change the color of the text, scroll using the Digital Crown while on the Color screen. To change which app data is displayed, tap the app widget you want to edit and use the Digital Crown to scroll through your options.

Motion

Compared to its some of its fellow styles, Motion is one of the more pared-down Watch Face. It includes the time and date, and the date's format can be adjusted or it can be turned off altogether, but other than that, Motion doesn't display any of your apps. Instead, you can choose between three lovely lightly animated image sets – butterflies, flowers and jellyfish (we're partial to the jellyfish, personally!). In the customization screen, use the Digital Crown to choose your image set. To see the next image in the set while the Watch Face is displayed, tap the screen lightly.

Simple

The Simple Watch Face is a bit more classic than Modular, though you can fit in almost as much app data. In Simple, you can adjust how much detail the clock shows, from a bare clock face with nothing but the hour, minute and second hands to a face calibrated down to the second. You can also adjust the color of the second hand by swiping to the Color screen and using the Digital Crown to scroll through the color choices. Finally, Simple includes five locations for app info. Tap them and use the Digital Crown to select your preferences.

Solar

The Solar style shows you the sun's current position in its daily trajectory. Like Astronomy, this style can't be customized, but it's neat to look at. You can use the Digital Crown to move the sun around and discover at what time solar noon, solar midnight, twilight, sunrise and sunset will occur / have occurred.

Unlike styles with app widgets, the Astronomy face is what-you-see-is-what-you-get. There are no customization options available.

Color

The Color face is very similar to the Simple face. It includes a customizable color clock face and four app widgets. It also includes a monogram area in the middle of the clock, which can be turned on and off.

Utility

The Utility style is a no-nonsense watch presentation. It includes three customizable widgets and an optional date stamp in the middle of the clock. You can also adjust the level of detail on the clock face and the color of the second hand.

Astronomy

The Astronomy face is one of our favorites, and it's sure to appeal to space and science lovers. This Watch Face gives you three different views – the Earth (and your current location on it), the moon in its current phase, and the planets in their current positions around the sun. For all three Astronomy features, you can use the Digital Crown to move forward and backward in time. It's very cool and fun to show off!

Chronograph

The Chronograph style is useful for people who need to use their Watch as a stopwatch regularly. It's designed to measure time extremely precisely, and includes a start button for stopwatch functionality. In addition, it can be customized to include four app widgets and the color of the stopwatch face can be altered.

X-Large

The X-Large style is the most minimal style of the bunch. It displays nothing but the current time. You can change its color by Force Touching the display, tapping **Customize,** and then using the Digital Crown to find the perfect hue, but that's about it. This is a great choice for users who like to keep things as simple as possible.

Mickey

The Mickey Watch Face style should be familiar to kids and former kids alike! Unlike a traditional Mickey Mouse watch, though, the Apple Watch version of this classic includes three customizable widget areas and an animated Mouse.

Saving Customized Watch Faces

As you can see, you can customize your Watch Face in hundreds of different ways. When you customize a style, it will retain your customizations, even if you switch to a different style. If you'd like to save a customization and create another Watch Face based on the same style, Force Touch the Watch Face display and swipe all the way to the end of the list. Here you'll find a plus sign labeled **New.**

Tap it and then use the Digital Crown to find the style you'd like to base your customization on. From there, make all the alterations you need. Your new Watch Face will be retained in your Watch Face style line up. Unfortunately, you currently do not have the ability to rename your creation, so you'll find multiple entries for the base style. If your list starts getting too unruly, you can delete Watch Faces by swiping up and then tapping **Delete.**

4.2 Notifications Management

We strongly – *strongly* – recommend getting a handle on your notifications. This can be the difference between a happy and productive Apple Watch user and an anxiety-ridden wreck. Okay, maybe that's a slight exaggeration, but a Watch that's constantly ringing, buzzing and thumping you may not be an ideal daily companion for anyone.

Most of your notifications will mirror the notifications on your iPhone, but they don't have to. To change Apple Watch notifications without affecting your iPhone settings, open up the Apple Watch iPhone app. Then tap **Notifications.**

Here, you can adjust some global Watch settings, including whether notifications are indicated with a red dot (recommended) and whether or not you'd like to turn on notifications privacy. This will prevent notifications from displaying any detail, saving you from embarrassing boardroom moments.

Beneath that, you'll see the nine apps that come pre-configured

to notify you – Activity, Calendar, Mail, Maps, Messages, Passbook & Apple Pay, Phone, Photos and Reminders. We know that your intentions are good, but the constant Activity notifications may start wearing on you, particularly the hourly Stand Reminders. You can turn those off, as well as progress updates and a few other options.

The other apps in this top list default to mirroring the settings on your iPhone. However, by tapping **Custom,** you can change their behavior on your Watch. You can turn visual alerts, sound and haptics (thumps) on and off here, and tell Watch whether or not you want it to repeat unread alerts, and if so, how many times.

Underneath these nine special cases, you'll find a list of every app on your iPhone that currently sends notifications. The apps in this list are going to send notifications to your Apple Watch too, unless you stop them! Simply use the sliders (green for on) to turn notifications for each app on or off.

Haptics Notifications

We were a bit nervous about the Apple Watch's haptic features at first. However, we found them to be much more gentle than we'd feared – too gentle, in fact. Fortunately for tough cookies like us, you can adjust the strength of your Watch's haptic feedback. Open the Settings app on your iPhone and tap **Sounds and Haptics.** There, you can adjust the strength of the haptic action, or turn on **Prominent Haptic,** which is much less ambiguous! You can also adjust these settings in the onboard Settings app on the Watch itself, but we found the sliders easier to manage on the iPhone.

4.3 Digital Touch and Friends Accessibility

Your Favorites are accessible by pressing the Side Button once. This brings up a wheel of up to twelve contacts of your choosing. Use the Digital Crown to scroll through your friend wheel.

You can add friends to this display through the iPhone Apple Watch app. Open it and then tap **Friends.** From there, tap **Add Friend** in each open slot and choose one of your preexisting iPhone contacts to fill the slot. You can also tap **Edit** in the top right corner of the screen to delete friends and rearrange their order.

You may find it useful to add yourself as one of your contacts.

This gives you the ability to send yourself iMessage memos through Siri.

Digital Touch

Apple Watch is a great tool for communicating with all of your contacts, but there are some extra communication features for your fellow Apple Watch owners.

You'll be able to tell which of your friends have an Apple Watch through the Friends screen. When you select an Apple Watch friend using the Digital Crown, you'll see the Digital Touch icon (a pointing finger) appear underneath your friend's picture. You can send your Apple Watch friends a sketch from here, which they'll receive as an animation. You can also send your heartbeat by pressing two fingers on the screen, or you can literally poke a friend by tapping the screen once (be careful with this – we have a feeling a little of this behavior can go a long way!).

4.4 Cashless transactions: the Apple Pay

Apple Pay on the Apple Watch completely eliminates the purse fumble at the grocery store by making it possible for you to use your Watch as a payment method, provided your bank supports it. You can find a regularly updated list of supported banks at https://www.apple.com/apple-pay/.

You'll need to set up Apple Pay on your iPhone before you can use it on the Watch. To do so, open the Passbook app on your iPhone. Follow the directions to enter a new card, either by using the camera or by entering the card number and expiration date manually.

You can also set up Apple Pay in the Apple Watch iPhone app. Just tap **Passbook & Apple Pay**, and then tap **Add Credit or Debit Card.**

In order to use Apple Pay with Apple Watch, you must set up a Watch passcode. You can do this in the Apple Watch iPhone app by tapping **Passcode.**

Once that's done, you're all set to shop. When you're in a store that can take Apple Pay payments, simply double tap the Side Button to open Apple Pay on your Apple Watch. Then move the face of the Watch so that it lines up with the Apple Pay reader (note that your Watch doesn't need to actually touch the reader). Not every store you visit will be equipped to accept payment through this method, but the numbers that do are steadily growing!

4.5 Customizable Apple Watch Bands

You can switch out Apple Watch bands fairly easily by pressing the small silver buttons on the back of the Watch itself. Press each button and slide off each half of the band you want to replace. Reverse the process to put a new band on. You might choose a classy leather band paired with the Simple Watch Face for a night out after your afternoon run with a sport band and the Chronograph Watch Face, for example.

4.6 Change the look depending on your mood!

One of our favorite things about Apple Watch is how quick its customizations are. It's a snap to change the entire look of your Watch depending on your mood – only to revert back to an earlier configuration within seconds later on. Apple Watch can easily keep pace with your changing moods, environments and needs. In fact, it's specifically designed to! Take advantage your options, whenever and however often you need to.

Next up, we'll learn how to use all of your Watch's onboard apps!

Chapter 5: Tweaking Apple Watch's Built-in Apps

Now that you have a pretty good idea of where things are and your Apple Watch is starting to feel a little more personal, it's time to get more specific about what it can do for you. In this chapter, we're going to cover every onboard Watch app and Glance that's built in to your Watch. When you've finished this chapter, you'll be able to truly consider yourself an Apple Watch expert.

5.1 The Native Apps

Your Watch includes nineteen native Watch apps. These nineteen apps can't be uninstalled or hidden, but we don't think you'd want to get rid of many of them. These apps make up your Home screen, which you can get to at any time by pressing the Digital Crown. We'll give you a brief overview of all nineteen here.

Activity

The Activity app is one of the most exciting features of the Apple Watch. It takes square aim at fitness wearable competitors like FitBit and Pebble by providing passive and active tracking, as well as health metrics and personalized goals. The first time you open Activity, you'll be prompted to enter your personal information to help the app give you accurate calorie burn estimates. Activity needs your age, gender, height and weight to function best (if you've already entered this information in Health, then Activity will bring it over from your iPhone for you). Next, you'll follow Activity's prompts to set your goals.

Activity tracks three metrics, nicknamed **Move**, **Exercise** and **Stand**. You'll be prompted to set goals for each, and you can (and should) adjust them as needed. The main Activity screen shows your progress toward all three goals by showing three circles. Scroll down to see your daily step count, active calories and distance travelled.

From the main Activity screen, swipe left to see detailed information about each of your goals. In each goal screen, you can swipe down to see a graph of your daily activity.

To adjust your calorie-based Move goal, Force Touch any Activity screen. Note that Activity automatically assigns Stand (one minute every hour) and Exercise (thirty minutes every day) goals for you.

Calendar

The Apple Watch Calendar app is an extension of its iPhone equivalent that allows you to see the current date and your upcoming calendar entries quickly and easily. You can get to the Calendar from your Watch Face if it's enabled there and from its related Glance as well.

Calendar will open to an event summary view, but to see the full month, just tap the date in the top left corner of the display. You can also Force Touch the display to see the option to get a day planner view.

Since Apple Watch doesn't include an onscreen keyboard, you can't create events on your Watch by typing. You can, however, enlist Siri for creating and editing events. Say, "Hey Siri, create calendar event titled [name] on [date] at [location]". Siri will then give you a chance to confirm that she heard everything correctly. If everything looks good, tap **Confirm** and your Calendar will be updated on both your iPhone and your Apple Watch.

Apple Watch will also notify you of incoming Calendar invites. You can respond to them directly from the Apple Watch notification.

Siri is great at entering simple events, but we do recommend using your iPhone or another Apple device to enter more complicated information, like location, event organizer contact information, notes and images. For more complex events, Apple Watch can help you stay on top of your schedule by giving you "Leave Now" reminders and helping you find directions to an event.

To set a Leave Now alert, you'll need to be sure that **Travel Time** is enabled in the event's details on your iPhone. When you turn on Travel Time, you can either set a time from the list or base it on a certain location. When that's configured, your Watch will alert you when it's time to go.

To get directions to an event, open the Calendar app and Force Touch the event. A button labeled **Directions** will appear. Tap that, and the Maps app will launch and load the directions from your current location to that of your event, and you'll be good to go!

Camera Remote

The Camera Remote Watch app functions as a remote control for the camera on your iPhone. It's a great way to take a picture without shaking your iPhone and blurring your image or to shoot a hands-free selfie or a group shot that includes you!

To use the Camera Remote app, first open the Camera app on your iPhone and set up your shot (if the Camera app isn't open, your Apple Watch will launch it on your iPhone for you). Then, use the shutter button on your Apple Watch to take the picture, or use the shutter delay for a three-second delay. Using the delay will cause your iPhone to take a burst of shots, so you'll definitely get at least one good one!

Photos you take using Camera Remote will be stored on your iPhone, but you can review your most recent shot by tapping the lower left corner of the Watch screen after you shoot. However, you'll lose this preview as soon as you close the Camera Remote app.

Alarm

Since Apple Watch will most likely need to be charged every night, you probably won't use the Alarm app as a morning alarm very often. However, the subtle haptic alarms of Apple Watch can still serve as useful reminders or gentle nap enders.

To add an Alarm, open the Alarm app from the Home screen. Then, Force Touch the display and tap the **New** icon. From there, you can adjust your alarm's time, reoccurrence and whether or not you'll be able to hit **Snooze** on it. You can even use the Alarm app as a substitute for your iPhone's Reminders app by giving your alarms labels. Note that in order to change the time of your alarm, you'll use the Digital Crown to scroll to the correct time.

Mail

We found that we did most of our email reading through the Notifications screen as new messages came in, but Apple Watch's Mail app provides access to your full backlog, if you ever need it. At the time of writing, you can't reply to messages in Mail, but you can take advantage of Handoff to open the message in Mail on your iPhone and reply there.

Mail on Apple Watch is fairly no-frills. You can perform some basic email management, though. When you're reading a message, you can Force Touch the display to reveal the options to flag or delete/archive a message or mark it as unread. Note that the delete/archive setting can be configured in the Mail app Settings on your iPhone.

Bear in mind that unlike its iPhone cousin, the Apple Watch is *not* designed to be a reading device. It's great for short messages, but email with lots of multimedia will not display perfectly on your Watch and you may want to read these sorts of messages on your iPhone instead. Handoff will make this much easier!

Mail Settings

If you get a lot of email, you may quickly tire of being thumped every few seconds. We highly recommend customizing your Mail settings to help make your Apple Watch a useful way to stay in the loop, as opposed to a constant stream of irritation!

To manage Mail settings, open the Apple Watch app on your iPhone. Then, tap Mail.

By default, your Watch is set to mirror the settings on your iPhone. However, if you tap **Custom,** you can change settings on your Watch without affecting Mail notifications on your iPhone. Here, you can turn alerts off for specific accounts (like that throwaway account you use for online shopping, perhaps). If you're drowning in email, we highly recommend setting up a VIP list consisting of friends and colleagues and then only receiving alerts from that list.

Music

The Music app on your Apple Watch is your onboard music player. Most of the time, you'll probably be streaming music from your iPhone, but it is possible to load a modestly sized music library directly onto your Apple Watch.

Music's most simple function is as a playback remote control for your iPhone. When you open Music on your Watch, you'll see the entire Music library of the iPhone you've paired with the Watch. You can browse your music by artist, album, song or playlist.

Tap on a song to play it. Note that playback will occur on your iPhone. When a song starts playing, Apple Watch will display the Now Playing screen. Force Touch the display here to reveal additional options, including Repeat, Shuffle, Source and AirPlay.

The **Source** icon lets you switch between music saved on your iPhone and music on your Apple Watch (more on this in a minute). The **AirPlay** icon will let you send audio to any AirPlay device you may have.

If you have Bluetooth headphones and don't feel like taking your iPhone along for your walk or run, you can save music on your Apple Watch and listen to it any time, regardless of whether or not it's connected to your iPhone. To add music, open the Apple Watch iPhone app. Tap **Synced Playlist.** Next, choose a playlist to sync with your Apple Watch. Note that you can control how much of a playlist to sync using the **Playlist Limit** on the Music settings screen. Apple Watch can hold up to 2 GB of music, which comes out to approximately 200 songs.

At the time of writing, you can only sync one playlist. You might consider creating a special Apple Watch playlist specifically for this purpose.

Once you've synced a playlist, you'll need to pair your headphones to enjoy your tunes, since your Watch's tiny speakers aren't quite up to the task. To do this, open the Settings app on your Watch and tap **Bluetooth.** Wait for the Watch to discover your headphones, and then tap them to pair.

Maps

The Maps app on your Apple Watch can help you figure out where you are and how to get where you want to go. When you open this app, you'll see your current location represented by a blue dot. You can use the Digital Crown to zoom in and out as needed.

To search for other locations, Force Touch the display. You'll then have the option to search for an address or look up a contact's address.

If you tap **Search**, you'll be able to search by speaking an

address or by scrolling through a list of recent Maps searches on your iPhone. One handy trick we picked up was searching for an address on our iPhone first and then opening the Maps Watch app and getting the location through the Recent list.

Searches for public places may include lots of other information, like a phone number, star reviews and hours of operation. Scroll down with your finger or the Digital Crown to see all of this information. You can tap phone numbers to place calls. You'll find directions listed here too, just above the map.

When you've requested directions, Apple Watch will load your route. Tap **Start** to begin your travels. Of course, you can always ask Siri for help. Just say, "Hey Siri, directions to [address]."

Once you've entered a route and started directions, Apple Watch will help you navigate through the display and through its Taptic Engine. Twelve steady taps mean that you have a right turn coming up. Three pairs of taps mean that you'll turn left. We promise that you'll get the hang of it quickly.

Once you arrive, Apple Watch will vibrate to let you know that you're there. If you need to exit directions before then, just Force Touch the display and tap **Stop.**

You can also use the Maps app to find an address of a location. Press and hold the screen until a purple pin drops on the map. Then tap the pin to see the address of its location, as shown below.

The Passbook app mirrors its iPhone counterpart by providing a

place to store boarding passes, tickets and more. We've had the best luck with Passbook in airports, which seem to have adopted onscreen boarding passes almost universally. There are other very practical uses for it, though, including the Starbucks Card, Fandango movie tickets, and more.

Passbook works by loading all of your passes in the iPhone Passbook app and then syncing them with your Watch. If you thought having a boarding pass on your iPhone was convenient, the wrist-worn version is even more accessible! Adding passes to Passbook usually requires the use of a Passbook-compatible app, like the Starbucks, Fandango, Airbnb, American Airlines, Eventbrite, Target and Walgreens apps (and plenty more). Purchasing tickets through those apps will result in Passbook-friendly passes. You can find a full list of Passbook-compatible apps by opening Passbook on your iPhone and then tapping the plus sign in the **Passes** section of the screen. This will give you the option to tap **Find Apps for Passbook.**

The Passbook app on your iPhone also stores your credit cards for Apple Pay purposes. You can add additional Apple Pay cards through Passbook or through the Apple Watch app. Just remember that you will need to add a passcode to your Watch in order to make Apple Pay purchases.

Phone

If you want to feel like the future is now, just make or take a call on your wristwatch. The Phone app on Apple Watch lets you make and answer calls through your iPhone, and it's awfully

satisfying to do so! Of course, if you ever need or prefer to take a call on your iPhone, you can swipe up on your Watch to answer on your phone instead.

 When you first open the Phone app on your Watch, you'll find your Favorites, Recents, Contacts and Voice Mail. All of this information comes from your paired iPhone.

Tap a menu item to open it. Note that you can listen to and delete your voice mail messages without having to touch your phone!

If you ever need to quickly silence a ringing Watch, simply cover the display with your hand. This will temporarily silence the device.

Like Mail, Messages may get the most use in the Notifications screen, especially since unlike Mail, you can actually respond to Messages directly from your Apple Watch! When you receive a text message on your iPhone (SMS or iMessage), your Watch will buzz, unless you ask it not to and a notification will appear. Just tap **Reply** underneath the message to respond. Alternatively, you can open the Messages app from your Home screen to see every message you've received on Apple Watch. Tap on a conversation to view it.

You have a few options for responding to a message on your Apple Watch. You can use some generic canned responses (scroll down to see the entire list), or you can respond with animated emojis or any of the standard iOS emojis. Tap the smiley face to enter the emoji menu. Swipe left and right to cycle through the basic emoji categories (animated faces, animated hearts, animated gestures and standard emojis). On each screen, you can use the Digital Crown to see every option available in each category.

If you'd like to add your own default Messages responses, open the Apple Watch app on your iPhone. Tap on **Messages**. Then, tap **Default Replies.** Type your new default reply (or replies) in the blanks.

You can also use dictation to send a message (and we recommend this, because it *definitely* feels like *Star Trek*!). Tap the microphone and speak your message into the Watch. You can send your message as an audio message if you prefer, or as a normal text message. Just tap either the audio recording or the text message to send it. You can also use your Apple Watch to send your location to fellow Apple enthusiasts. Force Touch the display in a Messages conversation to reveal the option to reply, see contact details, and send your current location in the form of a map. You can delete conversations by swiping left, and you can see when individual messages were sent by swiping right.

The Photos Watch app is similar to the Music Watch app. Nothing will appear here unless you sync it using the Apple Watch app on your iPhone. To do this, open up the iPhone Watch app and tap **Photos.** In the Photos settings screen, you'll find the Photo Syncing heading. Tap the **Synced Album** text to set up a new synced album.

You can only sync one album at a time, but any photos in a synced album are stored on your Apple Watch, so you can enjoy

them with or without your iPhone nearby. When selecting a photo album to sync, keep in mind that the Apple Watch is somewhat limited in storage space and will let you store up to 75 MB in Photos, which translates to about 500 pictures.

Your synced photos will appear in the Photos app in a mosaic. Use the Digital Crown to zoom in for a closer look at any of them.

While you can't take Photos on the Watch itself, you *can* take screenshots, which will be saved in your iPhone Photos app. Press the Digital Crown and the Side Button simultaneously to capture whatever's currently on your Apple Watch's screen.

Stocks

The Apple Watch Stocks app mirrors the Stocks app on your iPhone, meaning it will show you any and all stocks you follow on your phone. Use the Digital Crown to scroll through your list of stocks, and tap on any stock name to see details about it.

Settings

The onboard Settings app contains some – but not all – of the settings found in its iPhone counterpart. However, there is one setting you can only manage on the Watch itself. If you'd like to set your Watch ahead in an attempt to panic yourself into arriving on time, you can do so by tapping **Time**. All other settings described here can also be accessed on your iPhone's Watch app.

You can turn Airport Mode, Bluetooth and Do Not Disturb Mode on and off from the Settings app, though we've found it easier to use the Settings Glance to manage these common settings.

The **General** heading in the Watch's settings includes some useful information. Under **About,** you'll find your Watch's name, version, serial number, MAC address and more. You can also check your available storage space here and see how many songs, photos and apps are currently stored on your Watch.

Under **Orientation,** you can change wrists and Digital Crown sides. Under **Activate on Wrist Raise,** you can enable/disable your Watch waking up when you raise your wrist. You can also scroll down on this screen to choose whether your Watch will display the Watch Face or your most recently used app when you wake up your Watch.

The **Accessibility** settings available onboard the Watch include VoiceOver, Zoom, Reduce Motion and On/Off labels. Under **Siri,** you turn the "Hey Siri" command on and off. The **Regulatory** heading includes some required FCC information.

Finally, you'll find **Reset** at the very bottom of the Settings screen. Resetting your Apple Watch returns it to its factory default state and removes all of your personal information, content and settings. It's permanent, but since Apple Watch is so heavily dependent on your iPhone for content anyway, it's not a terrible way to start over from the beginning should you feel the urge. You will definitely want to do this before you sell or give away your Apple Watch (not that you'll ever want to!).

The main Settings app menu also includes the ability to manage your Watch's brightness and text size. Keep in mind that more brightness requires more battery! You can also enable bold text here to make your Watch a little easier to read. Under **Sounds & Haptics,** you can manage the volume of audio alerts or mute them altogether. You can also decide how forcefully you'd like the

Taptic Engine to nudge you when notifications come in. Finally, under **Passcode,** you can enable, disable or change your passcode.

Remote

The Remote app can be used to control iTunes on a Mac or PC, or to control an Apple TV. In order for this to work, your iPhone and computer and/or Apple TV need to be connected to the same wireless network.

To get started, you'll need to pair your devices. Open the Remote

app on Apple Watch and tap the plus sign to add a device.

Your Apple Watch will then display a four-digit passcode. To set up a connection with iTunes, click the **Remote** icon that appears in the top left corner of the iTunes screen (this icon will only appear if iTunes detects an Apple Watch running Remote on the same wireless network).

Then, enter the four-digit passcode from your Apple Watch.

iTunes will confirm that you can now control it using the Apple Watch Remote app.

Back on your Apple Watch, you'll now find your iTunes library in the Remote app. Tap the music note to control it.

Follow the same steps to pair your Watch with an Apple TV. While you can't do much with iTunes apart from playing, pausing, advancing and moving backward through the active playlist, you can actually swipe left, right, up and down on your Apple Watch to move through Apple TV's menu structure.

Stopwatch

The Stopwatch app is a handy way to time events. To use it in its default presentation, open it up and tap the green start button in the lower left corner. Tap the white button in the lower right corner to record laps. When you're done timing, tap the red button.

Once you're finished, you can swipe up to see your recorded laps.

You can also view the Stopwatch in Digital, Graph and Hybrid views by Force Touching the display. We found the Digital view to be the easiest to use, but you may have a different preference. Try them all out!

Weather

The Weather app mirrors your Weather app settings on your

iPhone, so if you've saved multiple locations on your iPhone, you'll be able to see them on your Apple Watch by swiping left and right.

The Weather app's default opening screen displays conditions in your current location for the next few hours. Tap the display to switch between temperature, cloud cover and precipitation forecast views. You can scroll down to see the ten-day forecast as well.

You can change your default weather city in the Apple Watch iPhone app. Tap **Weather** and then **Default City.** Choose the city from the list of cities currently associated with your iPhone's Weather app, or use your current location.

Timer

The Timer is a simple and useful app that behaves exactly as you'd expect. Open the app and enter the duration you'd like to time. Tap the hour box and/or the minute box and use the Digital Crown to enter values. If you need to set a timer for longer than twelve hours, Force Touch the display and then tap **24.**

World Clock

The World Clock app mirrors the World Clock app settings on your iPhone. It's a very useful app if you have family, friends or business in other time zones. You can add various international locations and figure out at a glance what time it is there.

To add locations to your World Clock, open the Clock app on your iPhone. Tap **World Clock** at the bottom of the screen, and then tap the plus sign in the top right corner to add locations. You'll choose from a fairly extensive list of major international cities representing all of the world's time zones.

Workout

The Workout app is designed for users who want to track serious cardio efforts. While the Activity app is a passive tracker that's very useful for making incremental lifestyle changes, Workout will help dedicated walkers, runners, cyclists, rowers and stair steppers keep track of their time, pace, mileage, heart rate and more. Start by choosing one of several available workouts – outdoor or indoor walking, running, cycling, elliptical machine, rower, stair stepper or "other" (in which Workouts will record calorie burn at the rate of a brisk walk).

Once you've chosen your workout type, you'll choose whether you want to base your workout on duration, distance or calories burned. After you've completed a few workouts, the app will suggest goals for you based on past performance. Of course, you can also choose an "open" workout, which will continue until you stop the tracker.

Tap **Start** when you're ready to go, and Workout will give you a count of three before it starts tracking.

As you work out, you can glance at your Apple Watch, which will display your elapsed time. You can swipe left and right to see more – distance, calories, etc. If you've started a workout with a time, distance or calorie goal, you'll see a progress ring that lets you know quickly how close you are to finishing.

When you're finished, Force Touch the display to end your workout. You'll see a summary that lets you know how far, fast and hard you went, and you can save or discard the workout.

To review your saved workouts, open the Activity app on your iPhone. Tap the date of your workout and then scroll down until you find the Workouts heading. Tap the workout (in the screenshot below, it's labeled "246 Calorie Outdoor Run") to see its details.

5.2 Apple Watch Glances

Now that you're familiar with all of your apps, let's turn our attention to Apple Watch Glances. Glances are based on the age-old action of glancing quickly at a wristwatch to check the time, and they're one of our favorite Apple Watch features. You can certainly glance at the time with Apple Watch, but you can also glance at the weather, your heart rate, stocks, your personal calendar and more through the magic of Glances.

Glances are like extensions of your Watch apps that display the most relevant information in a format that you can process in an

instant. By default, Apple Watch will display all of its included Glances – Now Playing, Heartbeat, Battery, Activity, Calendar, Weather, Maps, Stocks and World Clock. Many third party apps also include Glance views, which can be enabled on your iPhone.

To view your Glances, swipe up on your display. Then, swipe left and right to move from Glance to Glance. You'll find a handful of hidden features here as well, and so in this section, we'll cover each Glance on your Watch.

Managing Glances

Unlike native apps, every Glance can be hidden. To manage your Glances, open the Apple Watch app on your iPhone. In the **My Watch** screen, tap **Glances** about halfway down the screen. You'll then see two lists of available Glances. The top list shows the Glances currently installed on your Apple Watch. To remove any of them, tap the red circle to the left of the Glance's name. To reorder your Glances, use the three-line icon on the right to drag Glances up and down the list.

When you remove a Glance, it will move to the second list – the **Do Not**

Include list. Glances on this list can be quickly added to

your Apple Watch by tapping the green circle to the left of the Glance's name. Third party app Glances will appear here as well. Quite a few popular apps – Instagram, Twitter and Pandora, to name a few – come with their own Glance view, so be sure to check this list for extra Glances to add as you install new apps!

Settings Glance

One of the most useful Glances is the Settings Glance – here you can see if your iPhone is currently within range of your Apple Watch and quickly turn on/off handy features like Airplane Mode, Do Not Disturb and Silent Mode. Another very useful feature can be found here that allows you to play a sound on your iPhone. We seem to be constantly misplacing ours, and this locator feature is much easier to get to than digging out the iPad or MacBook and loading Find My iPhone! To play a sound on your missing phone, just tap the iPhone button at the bottom of the Settings Glance.

Heartbeat

Apple Watch's built-in heart rate monitor is one of its most exciting features, particularly if you're interested in improving your fitness. You can use the monitor through the Heartbeat Glance. When you open this Glance, it will begin measuring your heart rate. You can also see the last measurement taken and its time stamp.

The Heartbeat Glance isn't connected to a dedicated Watch app, but it *is* connected to your iPhone's Health app. Whether or not you use the Heartbeat Glance, your Watch will read your pulse every ten minutes. It then sends its readings to Health, giving you a stream of health data over time that might be useful for conversations with your physician or for tracking your personal fitness goals.

If you don't want your Apple Watch to track your heart rate or other fitness information, open up the Apple Watch iPhone app, tap **Privacy,** then **Motion & Fitness.** Slide the Heart Rate and/or Fitness Tracking sliders off.

Now Playing

The Now Playing Glance controls whatever audio your iPhone is playing (or has most recently paused). This includes the iPhone/Apple Watch Music app, but it also might include Podcasts, Pandora, YouTube, or any other audio source, whether or not it's available as a full Apple Watch app. Think of Now Playing as a sort of remote control for your iPhone's audio. You can advance or go back to a previous track, play/pause playback, and control volume.

Activity

The Activity Glance gives you a quick peak at how you're progressing toward your daily fitness goals. For us, it's easier to use the shortcut on our Watch Face to get to the full Activity app, but the Glance offers another way to see how things are going.

Calendar

The Calendar Glance lets you know what's happening today and tomorrow. Tapping the display will open the full Calendar app.

Weather

The Weather Glance uses your current location to let you know what's going on outside. The Glance will only show your current location, but tapping the display will open up the full Watch app where you can check on other saved locations.

Maps

The Maps Glance shows your current location. It was disappointingly slow to load for us, but once it does, it's a great way to get oriented in a new place.

Battery and Power Reserve

The Battery Glance lets you know how your Apple Watch is doing and how soon you'll need to charge it. The other really useful feature of this Glance is Power Reserve. Power Reserve shuts off every Apple Watch feature except for the time to help you prolong your battery life if needed. Once you enter Power Reserve mode, you can exit it at any time by pressing and holding the Side Button.

Tapping the Activity Glance will open the full Activity app.

Stocks

We're not entirely sure where Apple's commitment to its native Stocks app comes from, but if this is up your alley, you'll love this Glance!

World Clock

The World Clock Glance shows your first selected World Clock location.

Tapping the display will launch the app.

5.3 Perfecting Apple Watch's Apps and Glances

We've given you a *lot* of information in this chapter. If you're feeling a little overwhelmed, keep in mind that you don't have to use every single feature of your Watch, and in fact, you probably won't. Go ahead and trim down your Glances so that they only contain the information you really need and want. If an app doesn't do anything for you, don't worry about it! The best thing you can do to really get the hang of your Apple Watch is to identify its most useful features and practice using them. It's much easier to get used to swiping up for Glances and using the Digital Crown to find apps when the actions are connected to outcomes that are worth your time and attention.

At this point, we'd recommend taking a little time to get used to your native apps and Glances, because in the next chapter we're going to introduce you to the wide, wonderful world of third party apps!

Chapter 6: Onboard Third Party Apple Watch Apps

In this section, we'll help you get started adding third party apps to your Watch. Selecting and installing a great batch of apps is a critically important step in getting the most out of your Apple Watch. The native apps discussed in the previous chapter provide a great starting point, but it's the integration of the platforms and services you use everyday that really makes the Watch shine.

We'll start by explaining how to add third party apps, and then we'll share twenty of our favorites to help you get started!

6.1 How to Add Third Party Apps

Getting new Watch apps is a fairly simple process, but you'll need to do it all on your iPhone. When you download a Watch-ready app, you're actually downloading the iPhone version. The Watch extension is built-in. Similarly, if you already have apps on your iPhone that are listed as Watch-friendly, updating to the latest version will make the Watch extension available for you.

To add an app to your Watch, open the iPhone Apple Watch app. There, scroll down past the built-in Watch apps until you find a list of Watch-ready apps. You can tap on each one to toggle **Show App on Apple Watch.**

If you have a lot of apps and this seems too tedious, you can also turn on Automatic Downloads. This will automatically add new Apple Watch apps to your Watch as you download them from the App Store on your iPhone. To do this, open the iPhone Watch app. Then tap **General,** then **Automatic Downloads.** Turn on **Automatically Download Apps.**

6.2 The Best of the Best Apple Watch Apps

As the newest member of the Apple family, the Watch is only just beginning to amass its app stable. However, we expect it to

explode as developers catch up with all of this device's technological opportunities. That said, the apps available straight out of the gate are nothing to sneeze at! Read on for some of our favorites.

Social Apps

Being the new forefront medium of communication, the Apple Watch must stay true to its feature of being one of the most convenient way to stay up-to date socially. Here are our top-picks to keep you updated and always connected around the world.

Twitterific (Free+$1.99 IAP)

Probably one of iPhone Twitter's most visually appealing clients, Twitterific extends to the Apple Watch. With appealing notifications and a gorgeous interface, the Apple Watch is one of the must-have social apps for your wrist. It wisely omits a full feed, concentrating on interactions and your day's stats. Take note that you can also reply using Siri.

Instagram (Free)

The best of Instagram is now on your Apple Watch. You can browse your feed, like your favorite photos, and even leave emoji comments. And you'll stay up to date with interactive notifications sent directly to your wrist. You can use the Handoff to switch from your iPhone to your Watch.

Slack (Free)

The popularity of this app is unsurmountable, in terms of team communications, that is. Instead of scrolling through an entire feed on your wrist, you can just view direct messages and mentions that you can then reply with using succinct pre-defined answers, Emoji, or even by talking to Siri.

Skype (Free+ Network Fees)

First off, it lets you start chats on the go with your favorite contacts right on your Apple Watch. It also lets you receive notifications for new messages and reply to them straight from your wrist.

Skype for Apple Watch enables you to quickly and easily send messages using Siri-activated voice-to-text, emoticons, or predefined responses.

eBay (Free)

The wearable software both gives you alerts for auctions and lets you make quick bids -- you might prevent someone from sniping that antique auction without even reaching into your pocket. You can also reply to messages with voice dictation, and keep tabs on your top-level buying and selling activity.

Productivity Apps

The main purpose of a wearable is, well, to help you get more done. These apps will ease your everyday busyness and pack a punch with its efficiency, convenience, and usefulness.

Geofency Time Tracking (Free)

If your job requires you to track where you have been and for how long, there are a few apps that can log your time automatically when you leave and enter a geofenced location, making it easy to keep on top of the duration of your stay for, say, more accurate invoicing purposes. Geofencey Time Tracking lets you see recent destinations (with Map support) and hours spent in each location based on specific criteria, like the last day, week, or month.

Wunderlist (Free)

For collaboration, Wunderlist makes it easy to share lists or tasks of anything that needs to be done and you can even delegate tasks to to customized groups, whether it be family members, work colleagues, friends or some combo of the three. You create lists on your iPhone and iPad and access the lists on the Watch.

Evernote(Free)

If you're a fan of Evernote, you'll be happy to find you can dictate, view, and update notes on the go via dictation using the Watch. Of course, all new or modified notes will be synced back to all of your devices with Evernote installed, and you'll be notified by reminders and items that are due.

 If you're a frequent presenter who uses either of these apps, you should know that both Keynote and Powerpoint have been updated to allow control of your slides from the Watch. That means no more fumbling around for the office remote control when you're supposed to begin your talk.

If tracking finances from your wrist is useful, Mint lets you set and track daily and monthly spending limits, and lets you see how much you have spent for specific categories. Mint will also send notifications, which should help those who pay attention to it to stay on budget.

For wanna-be chefs and foodies, Epicurious lets you set a timer for more than 40 commonly cooked items like chicken, eggs, and pork right from your wrist. You'll get reminders of when to flip the items you're cooking, and the app promises to give pointers on what to look for when the food is properly cooked. Never burn a burger again!

 The Watch is best when used for quick interactions, and the Conversion app is useful because it allows you to quickly convert many common unit categories from your wrist. Categories include: length, temperature, weight, currency and time. You can even bookmark frequently accessed conversions for faster access.

If you're already a user of ProCamera, this Apple Watch app gives you a bit extra over Apple's own camera remote. Along with including a remote trigger, external viewfinder preview, photo preview and timer, you can configure the length of the timer delay and how many photos will be shot.

Weather and Travel Apps

Perfect for providing both information and update when you're out, the Apple Watch surely is a pinnacle for summarizing gist of information, especially when travelling. Here are our top apps for when travelling and being updated on weather.

TripAdvisor and FlightTrack: (Free and $4.99 respectively)

Frequent travelers should check out these apps from services like TripAdvisor, which lets you view your trip information, see reviews, ratings and even images of your destination. If you're more focused specifically on flight info, you'll want Flight Track 5, which delivers flight notifications, status, gate changes, as well as flight progress and related information.

Flush Toilet Finder Pro (Free)

There's nothing like wandering around a new city when nature calls; there's nothing worse than needing a bathroom and not knowing where the closest one is -- and speed is of the utmost importance. Flush Toilet Finder Pro gives you access to a database of over 100,000 public toilets, and offers directions on how to locate the closest one. Laugh all you want, but this app will help enormously in those moments when every second -- and every step -- counts.

Citymapper (Free)

If you're in one of the supported cities (which include Paris, New York and London), Citymapper is a must. It zeroes in on public transport, and provides precise, clear instructions on getting from place to place. You're informed about times for upcoming busses, trains or trams, and can access an outline of the stops to expect on your journey.

Weather Nerd ($3.99)

Powered by Dark Sky, Weather Nerd is in our opinion the superior option on Apple Watch. You get three panes to swipe between: Today, Hour and Week. Today shows temperature and rainfall estimates, along with a plain-English overview of the day's forecast; Hour details imminent rainfall; and Week offers an overview of what's coming during the next six days. (As for the 'nerd' part, that's left for iPhone, where the app contains a wealth of data.)

Yahoo Weather (Free)

If you fancy something a bit simpler for getting weather info, Yahoo's app fits the bill. It's all stylish icons and text with hints of glowing neon, and there's support for multiple locations. On scrolling, you'll coo at a beautifully animated daylight graphic, and tapping the current conditions loads a graph showing temperature, precipitation and wind estimates for the next several hours.

Find Near Me (Free)

This app is great for quickly finding nearby businesses by category, all by tapping a button: ATM; bank; bar; spa; zoo… As you can see, the odd option is a bit unusual, but the app's fast and even enables you to use your own search terms via Siri (currently a rare thing on Apple Watch). Individual items when selected usually provide additional details (addresses; maps) and reviews.

Health and Fitness Apps

For those extra—and we mean extra—conscious about their health and fitness, the Apple Watch can serve as your ultimate buddy. Together with that, we've gathered the top apps more fitting.

WebMD (Free)

If you're someone who has to take medicine regularly, the WebMD app offers info (via Glances you access by swiping up on the Watch face) and reminders about any medication you're taking along with details about when to take your meds, how much and how often. You'll never miss your water pill again. (See above.)

Fitness Spades (Free)

It's no secret that Apple is pushing Apple Watch as a fitness and activity tracker, and third parties are jumping on the bandwagon. Take for instance Fitness Spades, which makes working out a game. Press a button to draw a card, which correlates to a workout you can perform anywhere. The app tracks your activity and offers a helpful rest interval in between drawing the next card. The game is designed to push you to complete more cards in the deck than your previous time using the app.

VimoFit 7 Minute Workout ($1.99)

If you're into quick workouts, try the VimoFit 7 Minute Workout app. It guides you through a variety of exercises directly on the Watch.

RunKeeper (Free)

The popular iOS app RunKeeper has released an update that adds Watch support. Now you can start a workout, view stats, and end your run right from your wrist. And there's an option to shut out other notifications so you can focus on the road or path ahead without distraction.

Entertainment Apps

For some, the Apple Watch is more of a tool rather than a toy. However, that doesn't stop most from forming it as a new factor for boundaries set for fun. Here are our top recommendations for them.

Rules ($3.99)

Rules! gives you a daily mini-game challenge, which is all about memorising rules and tapping relevant cards. Easy! Only it isn't, because several rounds in, you'll be juggling a bunch of rules in your head ("Tap ascending"; "Reds if you see green"; "No animals"), which must be dealt with in reverse order, all the while knowing a single incorrect tap ends your game.

Shazam (Free)

There's still that sense of living in the future when it comes to Shazam. Waggle your phone about while a song plays in the background, and the app will reveal what it is. Now, you don't even have to get your phone out — just wave your arm around to reveal a song's title, and also lyrics, in case you want to leap on to the table and wow your friends with your vocal prowess.

Sky Guide (Free)

On iPhone, Sky Guide is the most beautiful and accurate of star and constellation guides. On Apple Watch, the companion app gives you a calendar of upcoming events, and optional notifications regarding what's about to occur in your location, for example so you can catch the International Space Station zooming overhead.

TuneIn Radio ($9.99)

TuneIn Radio provides access to over 100,000 radio stations from around the world. Using Apple Watch, you can change the current station your iPhone's blaring out, access recent and related stations, follow shows, and pause/play/skip. And if you don't fancy paying, the free version) works on Apple Watch as well.

Yelp (Free)

On your iPhone, Yelp offers access to over 70 million reviews of businesses worldwide. On Apple Watch, it focuses on filling your belly, giving you temptingly tappable buttons for listing nearby bars, restaurants and coffee shops. In each case, you can drill down into results lists or individual locations, perusing distances, maps, reviews and pricing indicators.

6.3 Exploring Further

When the Apple Watch debuted on April 24, 2015, there were hundreds of Watch-ready apps already waiting for it. That number has nowhere to go but up! You can always use the Featured screen on the Apple Watch iPhone app to see the latest and best-reviewed apps for Apple Watch. The Categories button in the top left corner of that screen can help you break things down as well. And don't forget about old-fashioned Internet searching – typing "Apple Watch apps" into a search engine will yield plenty of reputable sources, like CNET, 9to5Mac, Engadget, and more. Don't be shy about trying new things, especially if they're free!

We do firmly believe that patient Watch owners will be richly rewarded. Apple Watch is a very different animal, and we feel confident that the apps that take full advantage of its unique abilities and limitations are just around the corner. There are plenty of great apps available right now, but we also suspect that the best ones are still to come, so stay tuned!

Chapter 7: Apple Watch Alternatives

While the Apple Watch is the most commercially successful smart watch to date—even if it is just a couple of weeks old—ultimately, there are others who wouldn't prefer it. While most iPhone users prefer the Apple Watch as their primary smart watch accessory, there are others who seem to enjoy other alternatives due to several factors such as price, modifiability, etc. Of course, the Apple Watch isn't the first smart watch on the market—and it surely isn't the only one.

Surely Apple Watch is selling like pancake, which is why queues are longer than most. If your excited about having a smart watch to pair up with your iPhone, there are several alternatives out there that can also reach the bar of the Apple Watch.

Pebble

Offering a premium design, great battery life, easy to use interface, and a strong influx of apps, the Pebble is undoubtedly one strong competition against the Apple Watch. Like the Apple Watch, they can deliver messages, texts, e-mails, and notifications directly to your wrist. You can also play music with the watch and change the straps whenever you want. The Pebble time, just recently released, brings new software,a colored face, and a Timeline interface that can support voice replies, much like Apple's Siri.

Every Pebble watch also uses e-paper display which consumes far less power than the Apple Watch. They usually last a week or more before needing to recharged again.

Pros: Simple, sleek design. Pebble can be easily set up. Notifications are delivered without delay. The watch faces are customizable. It has a wide array of apps to choose from. It syncs well with both iOS and Android devices.

Cons: For some, the display size is too small. As of now, there had been some reported bugs for Pebble.

Martian Watches

Much like Apple Watches, Martian Watches are now capable of handling voice-commands. The response time of delivering notifications from your phone to your wrist is ultra quick. Martian Watches also come in varying styles, from square to round models, all coming in a variety of colors.

Pros: The battery life lasts longer as compared to other Apple Watch alternatives. Whenever you receive a call, the watch also displays the name of the caller. You can ignore calls by simply jiggling your wrist. The audio quality is also on point.

Cons: It doesn't have a full screen LCD but rather share the screen with an analogue clock on top of the display. It also isn't touchscreen which questions it's definition whether it truly is a "smart watch."

Microsoft

Even if it is a Microsoft product, it is iOS compatible. You'll get allows you to get email previews, calendar alerts, sleep track, and a 24-hour heart rate monitor. Guided workouts and GPS running maps are also available from the Microsoft smart watch.

Pros: Microsoft watch or rather Microsoft Band is voice activated, a software called Cortana handles all voice commands. It also has a lot of pre-installed apps you'll most likely find in any smart watch.

Cons: Third party apps aren't available for the Microsoft Band. It only comes with one design and style. Lastly, the Cortana feature is only applicable through Windows phones.

Alcatel One Touch

One notable feature of the Alcatel One Touch is that it's capable of supporting your music player. What's more, they can also track your physical activities, deliver notifications from your phone, and control your phone's camera. There are also several watch faces available for the Alcatel One Touch.

Pros: Impressive battery life that can outperform other alternatives. Compatible with both iOS and Android phones. There is also a built in USB charging port in the strap.

Cons: Some users reported that the watch is uncomfortably hefty. Third party apps aren't available. Connections are sometimes lost.

Garmin Watch

Yet another iPhone compatible watch, the Garmin Vivoactive brings the idea of tying the knot before smart phones and sports really close. Smartphone notifications, built-in GPS, and lifestyle tracking functions are readily available from the Garmin Watch.

Pros: Probably the first and only waterproof smart watch. The battery life also lasts longer than most. The watch also allows step tracking and is compatible with heart rate monitors.

Cons: It cannot be wirelessly synced. And only one profile can be synced with the watch.

Conclusion

We hope you've enjoyed getting to know your new Apple Watch. Now that you know how to navigate and customize your device, use all of its preloaded apps and features, add your favorite third party apps, and explore new possibilities in the App Store, you're all set. Whether you're traveling, working out, killing time, making plans or sharing with friends, your Apple